Math Concept Reader

City of the
FUTURE

by Ilse Ortabasi

Copyright © Gareth Stevens, Inc. All rights reserved.

Developed for Harcourt, Inc., by Gareth Stevens, Inc. This edition published by Harcourt, Inc., by agreement with Gareth Stevens, Inc. No part of this publication may be reproduced or transmitted in any form or by any means, electronic or mechanical, including photocopy, recording, or any information storage and retrieval system, without permission in writing from the copyright holder.

Requests for permission to make copies of any part of the work should be addressed to Permissions Department, Gareth Stevens, Inc., 330 West Olive Street, Suite 100, Milwaukee, Wisconsin 53212. Fax: 414-332-3567.

HARCOURT and the Harcourt Logo are trademarks of Harcourt, Inc., registered in the United States of America and/or other jurisdictions.

Printed in the United States of America

ISBN 13: 978-0-15-360256-6
ISBN 10: 0-15-360256-2

3 4 5 6 7 8 9 10 179 16 15 14 13 12 11 10 09 08

Harcourt
SCHOOL PUBLISHERS

Chapter 1:
Getting Ready

Mrs. Ng's students are looking forward to the annual City of the Future competition. A team of students represents Edison School at the competition each year. The students, with their teachers' help, design and construct a scale model of a futuristic city.

Not long ago, the Edison School team won first prize in the regional competition. They constructed their model city using containers and everyday materials such as plastic jugs, glass jars, and aluminum cans. They traveled to the national finals in Washington, DC.

Mrs. Ng tells her students that they will study three-dimensional solid figures before the competition. "Who can define a solid figure?" she asks the class. Huang says that it is a three-dimensional figure, such as a cube. It has length, width, and height.

Mrs. Ng draws a cube on the board. Then she writes *face* on one of the flat sides of the cube. She explains that the segment where two faces meet is called an edge. The vertex is the point where the edges meet. Mrs. Ng gives each student a sheet of construction paper. Each student draws a two-dimensional pattern on the paper, which can be cut and folded to make a cube. This two-dimensional figure is called a net. For homework the class must find the number of faces, edges, and vertices that a cube has.

Mrs. Ng announces at the end of class that students will build models from solid figures for the City of the Future competition this year. She encourages the students to think how spheres, cylinders, prisms, pyramids, and cones could be used to construct buildings for their city.

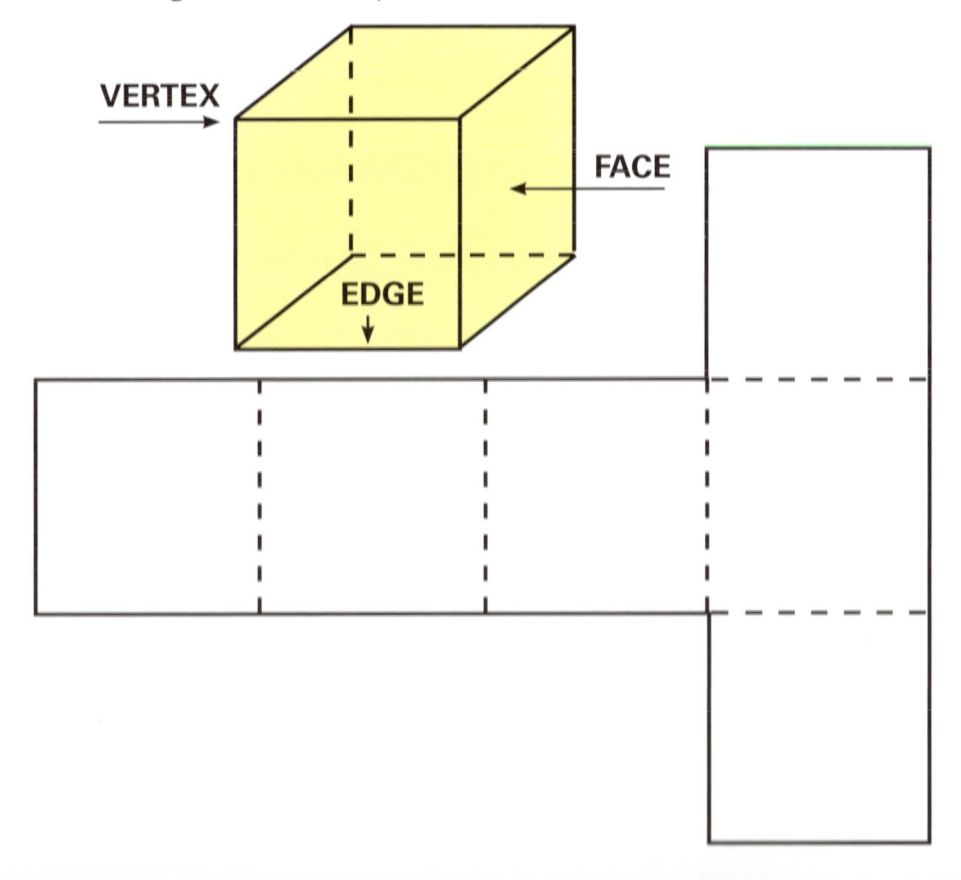

The Great Pyramid of Giza is an example of a square pyramid.

The class continues to study solid figures and their nets in great detail. Finally, Mrs. Ng announces that they are ready for the next step in their preparation for the City of the Future project.

Mrs. Ng sets up the computer and projector. After adjusting the lights and the blinds, Mrs. Ng shows the first photo to the class. It is a picture of the Great Pyramid of Giza in Egypt. Mrs. Ng says, "You see, even more than 5,000 years ago, people used solid figures to construct buildings." Mrs. Ng asks, "What type of solid figure is this?" "It is an example of a square pyramid," Zoe answers. "It has four triangular faces and a square base." "That's correct," says Mrs. Ng.

Mrs. Ng asks the students to find the definition of base in the glossary of their math book. Zoe is quick to find the answer. She reads, "Base: a solid figure's face by which the figure is measured or named." Mrs. Ng asks the students to find the definition of a square pyramid. Nathan reads, "A solid figure with a square base and with four triangular faces that have a common point." He adds that a square pyramid has five vertices and eight edges.

Huang wonders what the net for a square pyramid looks like. He draws in his math notebook while Mrs. Ng sketches the net for a square pyramid on the board. Mrs. Ng suggests that the students build a square pyramid out of construction paper. Soon everybody is busy with rulers, pencils, paper, scissors, and glue.

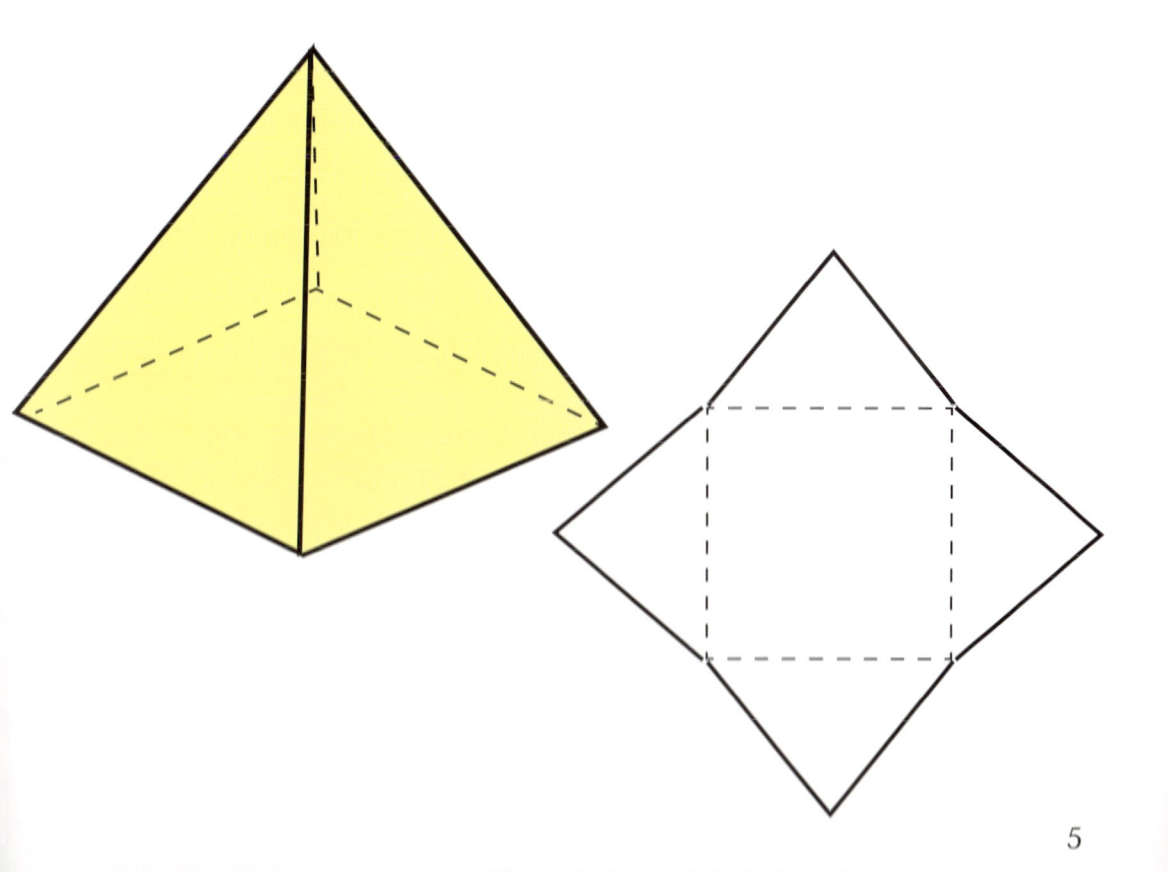

Chapter 2:
Buildings Shaped Like Solid Figures

    Students will construct solid figures to prepare for the competition. The nets help them identify the faces, edges, surfaces, and vertices of some of the solid figures. As they take the solid figures apart, they will be able to see which plane figures the solids are made from. The nets will also be patterns that Mrs. Ng's class can use to construct the figures they will use for the competition.

    Later, the class visits the media center. There they research existing buildings that are made from solid figures such as cubes, prisms, spheres, pyramids, and cylinders. The students are surprised to find so many interesting buildings in different places around the world. The media specialist helps the students search the Internet and books for images of buildings.

This cone-shaped building can be found in Spain.

Kendra finds an image of a cone-shaped building in Spain. It stands next to another building that has a unique design with attention-grabbing features. "What I like is that the two buildings are very different in shape. The cone is a good example of the solid figures we've been studying. The other building has curved features that make it stand apart from the cone shape. The differences make them complement each other. They go together in a way that makes each building interesting to see.

Kendra and Daniel decide to include a cone-shaped building in the design of the City of the Future. They are convinced that this piece will make their model city stand apart from the rest of the entries in the competition.

The Nasher Museum at Duke University houses a collection of modern and ancient art.

    Damon finds an aerial photo of a building that is made up of five rectangular prism-shaped buildings. The buildings meet at skewed, or slanted, angles at an entrance hall in the middle. "Looking at the buildings from overhead makes it easier to see the solid figures," Damon says. This is the Nasher Museum at Duke University in North Carolina. The museum has a collection of modern and ancient art.

    Damon and Natalie cut and fold the net for a rectangular prism. Each of the rectangular prism-shaped buildings has 6 rectangular faces, 12 edges, and 8 vertices. While they work on the nets, they decide if they want to have more than one rectangular prism-shaped building in the model city. "I think we should include five solid figures, so that it looks like the Nasher Museum," Natalie suggests.

This is a cubed-shaped apartment building in the Netherlands.

    Bob searches on the Internet for a cube-shaped building. He finds a photo of an apartment building in Rotterdam. Rotterdam is in the Netherlands. Bob shows the picture of the building to other students. They like the image because it has so many geometric shapes. The building has many cube-shaped boxes attached to the sides of the structure. The style and color of the cubes against the building makes it look like the solid figures are floating off the structure. Bob wonders, "What do you think it's like to stand in one of those cubes and look out the windows?"

    He constructs five cubes from their nets and glues them together. The result looks just like the cube-shaped parts of the apartment building he sees in the image.

This "house-on-a-pole" is located in Concord, California.

Emily finds a building shaped like a cylinder. In Concord, California, there is a "house-on-a-pole." She notices that there are two cylinders that form the building. There is one cylinder for the pole and another for the house. One cylinder is stacked on the other cylinder. "Look closely," Emily says. "To get into the house, you climb up a knotted rope or ladder." Emily wonders whether anyone lives in this house. It looks like it could be a great playhouse.

Cylinders have two circular ends, or bases, and a curved surface. Emily thinks about the kinds of materials she could use to build a cylinder for the City of the Future. She decides to build a cylinder from a paper towel roll. She uses two circular sheets of construction paper for the bases.

Chapter 3:

Soccer Balls and Solid Figures

Marcelo plans to play soccer after school with his friends. Right now they are finishing their homework during the after-school program. While he waits for his friends, Marcelo stares at the soccer ball. Marcelo thinks of the City of the Future project and what types of buildings the class will design.

Suddenly he realizes that the soccer ball is an interesting solid figure! It is similar to a sphere, but there are 12 faces on the ball shaped like pentagons. Marcelo counts 20 hexagon-shaped faces. Each pentagon is connected to 5 of the 20 hexagons. Marcelo thinks to himself. "One way to tell the difference between the hexagons and the pentagons is to look at the colors. On this soccer ball, the hexagons are yellow and the pentagons are blue."

Marcelo wonders what the net would look like if he could unfold the ball! Are there buildings that have this shape?

This geodesic dome has triangle-, pentagon-, and hexagon-shaped faces.

At school the next day, Marcelo asks Mrs. Ng about buildings with pentagons and hexagons as faces. He tells Mrs. Ng that he was thinking about this type of solid figure as he looked at his soccer ball. Marcelo's observation amazes Mrs. Ng. She says that an inventor, Buckminster Fuller, originally designed buildings shaped that way. These structures are called geodesic domes.

Later, Marcelo searches for a photo of a geodesic dome. He selects and prints a photo with the help of Mrs. Schulz, the media specialist. Mrs. Schulz says, "It's amazing how many geodesic dome images we were able to find."

As he looks at the photo, Marcelo realizes that he also sees faces that look like triangles. He decides to outline one of the hexagon-shaped faces. Marcelo will need to work hard to make a net for this solid figure!

The class has researched and studied many solid figures. However, they have not found a building shaped like a triangular pyramid. Daniel points out that a triangular pyramid is a different solid figure than the Great Pyramid of Giza in Egypt, which is a square pyramid. A triangular pyramid has a triangle as a base and three other faces that are shaped like triangles. It has 4 faces, 6 edges, and 4 vertices.

In order to model buildings to include in the City of the Future, Daniel goes to the media center. He and Mrs. Schulz work together. They find drawings of a triangular pyramid and its net. Daniel discovers that the net of an equilateral triangular pyramid is a triangle, too. It is four times the area of one of the faces of the pyramid.

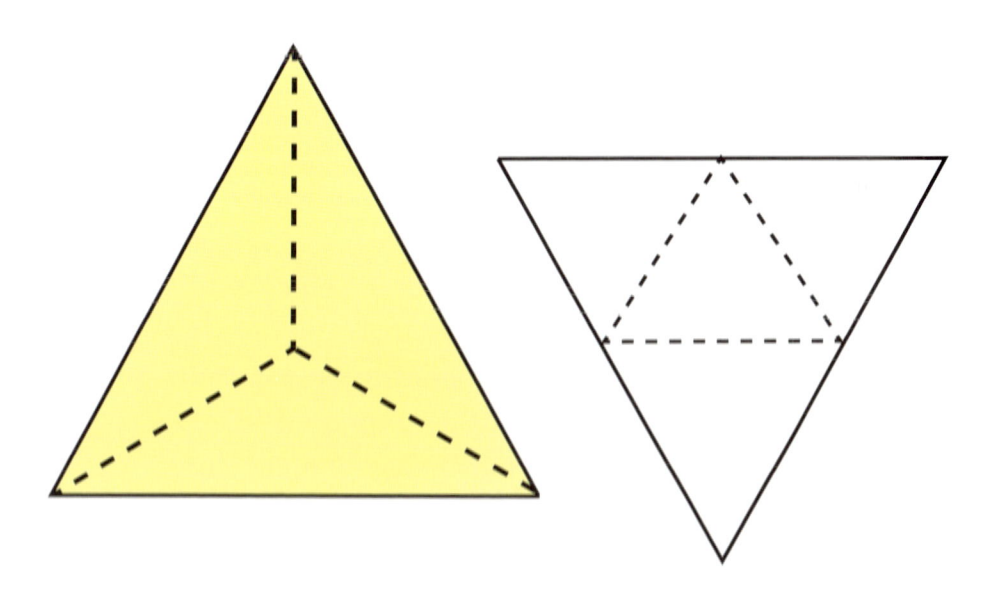

Mrs. Ng helps the students plan the residential part of the city.

    The classroom is covered with construction paper models of solid figures. Now the task is to build the City of the Future using all the shapes they have created. The students have many ideas. Mrs. Ng reminds them that they can make more shapes if they need them.

    The model city will be built on a large piece of plywood. Mrs. Ng divides the class into groups. One group will build the residential part of the city. Another group will build the commercial and industrial areas. The third group will build the transportation system and power plants. Emily reminds everyone not to forget recreational areas, like parks, sports fields, and swimming pools.

    As they get busy working, the students talk about the images they researched. These buildings inspire them to create an amazing model city.

City of the Future competition encourages students to think about city planning.

The students write a description of the city in their project plan. The description explains how everything will work when the futuristic city is built. The class constructs its model city. The students talk excitedly as they begin their work. There are many problems to solve and decisions to make as they build the city. The students agree that math is important in solving the problems of city planning.

Three students will represent the class during the regional competition and will explain the key design features of their city.

When the day of the competition arrives, everyone wishes the students good luck. The students carefully load their City of the Future model on the bus and then climb aboard to head for the regional competition.

Glossary

cone a solid figure that has a flat circular base and one vertex

cube a solid figure with six congruent square faces

cylinder a solid figure that has two parallel bases that are congruent circles

edge the line made where two or more faces of a solid figure meet

equilateral triangle a triangle with three congruent sides

face a polygon that is a flat surface of a solid figure

hexagon a polygon with six sides and six angles

net a two-dimensional pattern that can be folded into a three-dimensional polyhedron

pentagon a polygon with five sides and five angles

pyramid a solid figure with a polygon base and all other faces triangles that meet at a common vertex

rectangular having four right angles

scale model a smaller representation of something

sphere a round object whose curved surface is the same distance from the center to all its points

triangular having three angles

vertex the point where two or more rays meet. The vertex is also the point of intersection of three or more edges of a solid figure or the top point of a cone. The plural of vertex is vertices.

Photo Credits: cover, title page: Andy Caulfield/Riser/Getty Images; p. 4: © Roger Wood/CORBIS; p. 7: Davis McCardle/The Image Bank/Getty Images; p. 8: © Brad Feinknopf; p. 9: © Jon Hicks/Corbis; p. 10: Courtesy of Dave Michmerhuizen; p. 11: Photos.com; p. 12: © Joseph Sohm/Visions of America/Corbis; p. 14: © Jonathan Nourok/PhotoEdit Inc.; p. 15: Getty Images Entertainment/Getty Images